NUMBER 627

THE ENGLISH EXPERIENCE

ITS RECORD IN EARLY PRINTED BOOKS
PUBLISHED IN FACSIMILE

ROBERT TANNER

THE READY USE OF THE SPHERE

LONDON, 1592

DA CAPO PRESS
THEATRVM ORBIS TERRARVM LTD.
AMSTERDAM 1973 NEW YORK

The publishers acknowledge their gratitude to
the Curators of the Bodleian Library, Oxford
for their permission to reproduce the
Library's copy, Shelfmark: 8°.T.15 Art. BS

S.T.C.No. 23671

Collation: ∴ ^{8}, A-G^8

Published in 1973 by

Theatrum Orbis Terrarum Ltd.,
O.Z. Voorburgwal 85, Amsterdam

&

Da Capo Press Inc.
- a subsidiary of Plenum Publishing Corporation
277 West 17th Street, New York N.Y. 1011

Printed in the Netherlands

ISBN 90 221 0627 6

Library of Congress Catalog Card Number:
73-6164

j.

Anno Domini. 1592.

A briefe Treatise for the
ready vse of the Sphere: Lately made and finished in most ample large manner.

By Robert Tanner Gentleman,
Practicioner in Astronomie and Phisicke.

In the Which Globe or Sphere, there is added many strange Conclusions, as wel Cœlestiall as Terrestriall, the like heeretofore neuer deuised by any. Necessary not onely to those that follwe the Arte of Nauigation: But also to the furtherance of such as bee desirous to haue skill in the Mathematicall Disciplines.

Thou O Lord in the beginning hast laid the foundations of the earth: and the heauens are the woorkes of thy hands. Psalme. 102. ver. 25.

To the most high, mightie & renowmed Princesse, and most dread redoubted Soueraigne Ladie ELIZABETH, by the grace of God Queene of *England*, *Fraunce*, and *Ireland*, Defendresse of the faith.

MOST Excellent, gracious, and sacred Soueraigne, I was mooued to take in hande to compyle out vnto your Highnes (so well as my simple learning would

The Epistle

serue me) a briefe Treatise for the readie vse of the Spheare or Globe: A worke, by your poore obedient subiect, with great paines, labour, and studie, made and inuented: with sundrie additions and formes, not heeretofore deuised by anie: no lesse pleasant than profitable for the benefite of your Highnes Common-weale, to the studious in the Mathematicall Sciences, and to the furtheraunce of Trauellers in the Arte of Nauigation, and to all other your Highnes subiectes that are desirous of the knowledge of the beautifull frame of the Celestiall Orbes with their quantities, distances, courses, and strange intricate

Dedicatorie.

tricate miraculous motions of the resplendant Globes of the Sunne, Moone, Planets, and Starres fixed. And where it is a thing grafted in Nature, and naturall through custome (right renowmed Princesse) by manifolde sundrie meanes, either by the faculties of the mind, the qualities of the bodie, or the gifts of fortune, for men that are led on by inward affection, to seeke the friendship of those whom they affectionate, or rather inwardly loue: Wherein euerie man is so liberall, as the giftes of the minde, bodie, or fortune will affoord him. Wherein, although I am not so farre endued with anie of them,

The Epistle

them, that thereby I may deserue
anie thing at all: and although the
giftes of my minde are vnperfect,
if they bee placed against such an
obiect, (as without assintation bee
it spoken) your Maiesties Royall
selfe is; yet weighing with my self,
the bountifull goodnesse of your
Highnes nature to resemble a royall and fruitefull Tree, which the
more it is loaden, the more it decli-
neth: and the naturall gentlenesse
and royal clemencie of your Maiesties courtesie, wherewith your
Highnes is wont to receiue fauourable, to conceaue and iudge roy-
ally, of whatsoeuer your Maiesty
perceiueth, either to be ouer-slipt by
negli-

Dedicatorie.

negligence, let passe by infirmitie, or inserted by ignorance. And considering likewise the Prouerbe: Candidæ Musarum ianuæ, the doores of the learned are free from enuie: I purposed by these small fruites of my skill, to make attempt if I could win anie fauourable acceptance, within the clemencie of your Highnes fauourable protection. Perswading my selfe, that I might safely commit my rude work to the presence of your Royall Maiestie, who will wisely winke at my wants, and honorably construe the good meaning of my minde. And the more willing was I ledde on to seale it, and as my faithfull alleageance

The Epistle

ance bindeth me like a dutifull and obedient subiect, to chose out your Highnes most royall worthines aboue all other Princes, vnder the Orbes of the Heauens, to present this small worke of mine: because your Maiestie can learnedlie iudge of that which the ignorant can not comprehend. The onely name of your Highnes royall sacred Maiestie, shall be vnto me a sure target of steele, to beate backe the glauncing strokes of vnskilfull tongues: Who when they can not find out, by naturall reason, the quantitie and qualitie of superiour bodies, will vtterly condemn the rules of Arts, and quite ouerthrow al euident de-

mon-

Dedicatorie.

monstrations: against whom (because they denie principles) there is no reasoning nor disputing at all.

If I should stand vpon termes of Arte, and goe about to vnlock the closet of Astronomie vnto your Highnes, whom I know sufficient of your royal selfe to wade through the deepest flouds and swiftest streames contained in the Mathematikes: your Maiestie might iustly saye, Sus docet Mineruam; & I might rather discouer that I want, than your Highnes want that which I set downe. It is better therefore for me to bee silent, considering with what a wise, learned, royall Princesse I haue to deale, than by needlesse

The Epistle

lesse words to open mine owne defects: humbly crauing pardon of your Maiestie for this my bold enterprise hoping that your Highnes will not mislike this simple signifying of my bounden duetie. For like as the myte of the poore widdowe mentioned in the holie Scriptures, which she gaue in all her penurie, is accompted a greater gift than those huge summes that great men laid out of their great store: So this my rude Edition of my simple handy worke, if it may be accepted into your Highnes tution, it shal incourage mee heereafter to practise workes of greater importance; and (as it is my bounden duetie) al-

wa es

Dedicatorie.

waies to praye to the *Almightie God to blesse and keepe your Maiestie in health, tranquilitie & peace,* So my daylie prayer is to the same God, to blesse and defend your Highnes from all your enemies, and to raigne and liue ouer vs Nestors yeares.

Your Maiesties

most humble subiect,

Robert Tanner.

Anno Domini. 1592.

A briefe Treatise for the
ready vse of the Sphere: Lately made and finished in most ample large manner.

By Robert Tanner Gentleman,
Practicioner in Astronomie and Phisicke.

In the Which Globe or Sphere, there is added many strange Conclusions, as wel Cœlestiall as Terrestiall, the like heeretofore neuer deuised by any. Necessary not onely to those that follwe the Arte of Nauigation: But also to the furtherance of such as bee desirous to haue skill in the Mathematicall Disciplines.

❧ Thou O Lord in the beginning hast laid the foundations of the earth: and the heauens are the woorkes of thy hands. Psalme. 102. ver. 25.

And nowe to begin howe to erect the Sphere or Globe.

But first, the difinition what a Sphere is, with his partes, are to be known and found out.

A Sphere is a massie body, inclosed with one platforme; and in the middle of it ther is a pricke: from which, all lines drawne to the said platforme, are equall each to other, and that pricke is the Center of the Globe, and so sayth *Enclid*.

A 2 The

The Axeltree is a right lyne (which mooueth not) and passeth through the Center of the Globe, at which ends are imagined the Poles of the world; one is named the North Pole, the other the South.

The North Pole is called, *Pole Artick*, and the other Pole in the the South, *Antarticke*: the South Pole is neuer seene of vs in this our Countrey, but is euermore vnder our Horizonte. The North Pole is alwayes seene of vs whereas wee dwell, and these starres be opposed the one right against the other.

The

The Meridian is a great Circle passing ouer our heades, in which Circle when the Sunne is, hee maketh the middle of the day and the middle of the night. And also, in this Circle is shewed the Latitudes of all places, by the height of the Poles, in euery seuerall Latitude.

The Horizont What it is.

The Horizonte is a Circle, which goeth a-long by the edge of the ground, and parteth the part of the world which we see, from that part which we see not; and when the

Sunne

Sunne riseth, then he is in our Horizonte, and so is he when he is going downe as lowe as wee can see him.

Also it deuideth the whole Sphere of the world into two equal parts, in such sort, that halfe of the Sphere is euer aboue the ground, and halfe alwayes vnder the earth. This Circle hath great vse in the heauenly motions, that by it wee iudge the rysings & settings of the Sunne and Moone, and all other Starres; And in this Horizonte you shall finde noted, the names of the winds, which the Marriners vseth, by the 32. poynts of the Compas.

The

The next Circle is noted, the degrees which euery day in the moneth dooth contayn: that is to say, the first day one degree; the second day two degrees; and so forth as they succeed in order, to a point like a little starre, where the last day of euery moneth endeth.

The next Circle sheweth, what day of the moneth the Sunne entreth into any of the twelue signes, telling euery day one degree, to thirtie degrees, & so they succeed through euer one of the twelue signes, monethlie.

This *Horizonte* is deuided into foure quarters, East, West, South, and North, euery quarter of the world contayneth 90. degrees: and the whole compasse therof, is 360. degrees.

The next Circle is *Motus trepedationis*, a starry firmament, whose motion is slowe, from the West to the East, that euery hundreth yeere (by the obseruatiō of diuers Astronomers) moueth but one degree.

The *Zodiacke*, is a great broade and slope or shoring Circle, in the which are depictured the twelue signes and fixed starres, in the middest

middest whereof, is the Eclipticke lyne, from which the Sunne neuer swarueth.

Then followeth the two Colluers, and the Equinoctiall Circle, parting the Sphere in the verye middest, betwixt the two Poles: by reason whereof, there are two Latitudes, the one is North and the other South.

The North Latitude is contayned betwixt the Equinoctiall and the North Pole; the South Latitude, is betwixt the Equinoctiall and the South Pole: either of these two spaces contayneth in bredth 90. degrees.

A Degree is one part of a Circle beeing deuided into 360. parts, and 360. degrees, is the very Longitude of the Earth : and at the furtheſt Meridian in the Weſt, beginning with one degree, and ſo proceede Eaſtward, vnto 180. degrees of the Equinoctiall, & from thence goe forward to the Weſt, where you come againe to 360. degrees, which is the laſt degree of Longitude.

Next followeth the two tropicall Cyrcles.

That is to ſay, the North tropike
is

is *Cancer*, and is the returne of the
the Sunne in Sommer declyning,
backe againe towardes the Equinoctiall, the dayes being then at the
longest, and the nights at the shortest with vs, and then beginneth the
dayes to shorten againe.

The Winter Tropicke (sayth
Proclus) is the most Southerliest
Circle of all them that the Sunne
dooth describe, by the reuolution
of the world, in the which when
the Sunne is, he maketh his winter-
lie turne, and then is the longest
night in all the yeere and shortest
day with vs.

Para-

Paralels.

The Paralell lynes are described by the wyers in the Globe, and a Paralell of the longest day, is a space of the Earth: by thys is knowne the increase of the day to be a quarter of an hower, going from the Equinoctiall towards any of the Pole starres.

A Clymate contayneth two Paralels, in which spaces the day increaseth by halfe an hower: Of these Paralels are made 24. Climates, betweene the Equinoctiall and the tropicke of *Cancer*.

Then followeth the Artick Circle, and the Antartick Cyrcle.

The Artick Circle is the North Circle: and the contrary Circle in the South, is called the Antartick Circle: by the which Greeke composition, as you would say, contrary or against the Articke Circle, & it well may bee called the South Circle. But now heere how *Proclus* defineth them.

The Articke Circle is the greatest of all those Circles which doe alwayes appeare, and toucheth the Hori-

Horizonte in one only poynt, and is altogether aboue the Earth, and all the ſtarres that be within this Circle neyther riſe nor ſet, but are ſeene to runne round about the Pole all the night.

The Antarticke Circle is equall & equidiſtant to the Artick Circle & toucheth the Horizonte in one onely poynt, and is all vnder the ground, & all the ſtarres that be in it, are euermore out of our ſight.

Then is there two other ſmaller Circles, called Poller Circles, or Pole Circles: in this Circle about the Antartick Pole, is deuided certaine tayne degrees to take the Altitude of the North starre, this starre is in the extremitie, or end of the tayle of the lesse Beare, being a constellation, commonly called the Horne: for this North starre (of the most notablest starres aboue the Pole) is neerest vnto it, & shall therefore shewe a lesse Circle than any other, and so shall his Altitude differ little from the Altitude of the Pole.

This starre hath declination 85. degrees, and 51. minutes, and the complement of nintie (which are foure degrees and nine minutes) is his distance frō the Pole. And although

though the Marriners hold opinion, that it is not diſtant more than three degrees & a halfe, yet to the iudgmēt of thoſe perſons thathath knoweledge in Aſtronomie, more credite ought to be giuen to the Aſtronomer than to the Marriner, for aſmuch as the Aſtronomer doth know the place of the ſtarres, with their Longitudes, Latitudes, declinations, and right aſcentions, more perfectly and preciſely than dooth the Marriners: for they accompt not onely by degrees, but alſo by minutes and ſeconds; therfore whoſoeuer wil preciſely know it, let him take the higheſt Altitude of the North ſtarre, which is
his

his beeing ouer the Pole, and the leſſe Altitude, which is his beeing vnder the Pole: then take away the leſſe from the more, and the halfe of that remayneth, ſhalbe the diſtance of that ſtarre, from the Pole of the world. And likewiſe by thys experience may be known the Altitude of the Pole, and what all the other ſtarres that goe not downe vnder the Horizonte, be diſtant from it, ioyning the greater Altitude with the leſſe : and that ſhall amount thereof, deuided by the halfe, ſhall bee the Altitude of the Pole ; and taking awaye this Altitude of the Pole, from the greater Altitude of the Starre, or the leſſe from the Altitude of the Pole, the rest that remayneth, shal be the distance of the starre from the Pole. And as the Pole is inuisible, it can not be seene or known when the North starre is higher & lower, except it be by the meane of some other marke. And for this is considered, the position of the former Guardes or Watch, beeing one of the two starres called the Guardes, which are in the mouth of the Horne: the Marriners haue noted eyght positions, from the former Guard starre to the North starre, which aunswereth to the eyght principall windes; and as the Guarde is to the North starre, ac-

cor-

cording to the placing of these positions, so it shall be higher and lower from the Pole.

Let vs heere put the common Rules which the Marriners vse, to compile with those that are of opinion of three degrees and a halfe.

And for the opinion of Astronomers,(which is the distance of 4. degrees & 9. minutes) I haue in my Sphere or Globe annexed to my Diall in the North end, a Circuler or Figure with a moueable Horne; Vppon the vttermost Margent of the Diall, is noted the eyght winds of the eyght positions, and also the other points belonging to the Mariners Compas, and putting the Guardes and the North starre in euery of the Windes, it shall be the distance that the North starre is higher and lower than the Pole, as by the thrid appeareth in the cutting of the degrees in the Pole Circle, when the Horne is mooued too and fro.

Commmon Rules after the Marriners acompt, but not after the Astronomers acompt, as may appeare in the Rules next before.

THE former Guarde being in the East, the North starre is one

one degree and a halfe vnder the Pole.

The Guard being in the Northeast, the North starre is three degrees and a halfe vnder the Pole.

The Guarde being in the North, the starre is three degrees vnder the Pole.

The Guard being in the Northwest, the starre is halfe a degree vnder the Pole.

The *i*Guarde in the West, the starre is one degree and a halfe aboue the Pole.

The Guarde in the South-weſt, the ſtarre is three degrees and a halfe aboue the Pole.

The Guarde in the South-eaſt, the ſayd North ſtarre is halfe a degree aboue the Pole.

And thus in the Dyall and the Circle about the ſame, may you ſee the North ſtarre, in what part it is of the degrees, high or lowe, from the Pole: not onely by the Marriners Rule, but alſo by the Aſtronomers Rule.

And being thus knowen, howe
much

much the North starre is vnder &
aboue the Pole, let vs take the Altitude thereof.

And that of it that is vnder the Pole, let vs ioyne to his height, and as much of it as is aboue, let vs take away, and that shall rise therof, shall be the Altitude of the Pole aboue the Horizonte.

Thus much for breuitie sake, I haue borrowed and collected out of *Martine Cortese*, and other good Authors, these fewe notes, for the vse of the Guardes, fixed in the North end of my Sphere or Globe. And nowe I am to intreate of the inwarde part of the Cœlestiall

Globe, and also of the Terrestiall Globe; and to beginne with the lowest first.

The Terrestiall Globe hath depictured vpon it, a Mappe or Cart of the description of all the Earth, and the chiefest Regions, Citties, and Townes vnder *Europe*, *Affrica*, *Asia*, and *America* : And in thys Globe is contayned two Elements, that is to say, *Earth* and *Water*. The *Earth* is lowest of all Elements, black, ponderous, round, inuironed and inclosed with the other three; she is called the Mother of fruits, the roote of all plantes, the nourse of lyuing Creatures, the foundation
of

of buildings, the Sepulchre of the dead, the Center of the beautifull frame of the world, the matter and substance of mans body, and the Receptackle of heauenly influence: she is also garnished with fragrant flowers, with beautifull collours of Man, Beast, and Foule, inhabited, and comfortably quickned by the nourishing beames of the Sunne, Moone, Plannets, and fixed starres.

The Earth in comparison to the whole world, is but a pricke or mote, the whole compas thereof, is 360. degrees, & euery degree is 60. myles: and yee multiply 360. degrees by 60. it yeeldeth 21600.
miles

myles about the same.

The next Cyrcle aboue the Terrestiall Globe, is the Element of *Ayre*: and the next Region aboue the *Ayre*, is the Element of *Fyre*: and there are the foure Elements, described in this Sphere or Globe.

Then ensueth the Spheres of the 7. Plannets: that is to say, the first is the Sphere of the Moone: the second is the Sphere of *Mercury*: the third is the Sphere of *Venus*: the fourth is the Sphere of *Soll*: the fift is the Sphere of *Mars*: the sixt is the Sphere of *Jupiter*: the seauenth is the Sphere of *Saturnus*: the

the eyght is the Sphere of the starrie Firmament; and euery one of these Spheres dooth carrie hys signe and Caracter vpon him.

And according to the common accompt, the Earth is 39. times so much as the Moone. But the Sphere of the Moone is farre bigger than the Globe of the Moone, & the semidiamiter of her Sphere, is 33. times $\frac{1}{2}$. longer than the Earthes semidiamiter, & the myles of the semidiamiter of her Sphere is 115278. and the myles of her Sphere in compasse, contayneth 724604. $\frac{4}{7}$.

The semidiamiter of the Sphere of

of *Mercury*, is 64. times so long as the Earthes semidiamiter : the miles of the semidiameter, containes 220500. $\frac{2}{33}$. And the myles of his Sphere in compasse, contayneth 1386000. $\frac{4}{131}$.

The semidiamiter of the Sphere of *Venus*, is 167. tymes so long as the Earthes semidiamiter: the miles of the semidiamiter, contayneth 573872. $\frac{3}{11}$. the myles of the Sphere in compas, contayneth 3607200.

The semidiamiter of the Sphere of the Sunne, is 1120. tymes so long as the Earthes semidiamiter: the myles of the semidiamiter, contayneth

neth 3848367. 3/12 the myles of hys Sphere in compasse, contayneth 34189737. 1/7.

The semidiamiter of the Sphere of *Mars*, is 1220. times so long as the Earthes semidiamiter: the myles of the semidiamiter, contayneth 4192363. 7/11. the miles of the Sphere in compasse, contayneth 26352000.

The semidiamiter of the Sphere of *Jupiter*, is 8876. times as long as the Earths semidiamiter: the myles that the semidiamiter contayneth, is 30501163. 7/11. the myles of the Sphere in compasse, contayneth 191721600.

191721600.

The semidiameter of the Sphere of *Saturne*, is 14405. times so long as the Earthes semidiameter: the myles that this semidiameter contayneth, is 4950318. $\frac{2}{11}$. the myles of this Sphere in compas, contayneth 311148000.

The semidiameter of the eyght Sphere, is 20110. times so long as the Earthes semidiameter: the myles that this semidiameter contayneth, is 69105272. $\frac{1\cdot1}{8}$: the miles of this Sphere in compasse, containeth 434376000.

In

In this Armill or Ring Sphere, are wonderfull conclusions to bee learned, very strange and maruellous to the simple & ignorant persons, voyd of thys knowledge, no lesse profitable than commendable to them, and to the skilfull & wise; for the vse thereof is very apt and ready in teaching, and is more easie for young learners, than the Sollid or Massie Globe. And this is a maruellous excellencie in knowledge, to bee able so certaynly to iudge of things absent, as if they were present, to be able to tel what hower of the day it is in all parts of the Earth, and when the Sunne ryseth and setteth in all places vnder

hea-

heauen: for the howers of the day
are dyuers in dyuers Regions; so is
the shaddowes that the Sunne
causeth in their Dyalls, and all o-
ther shaddowes doth disagree many wayes, not onely from our shaddowes, but also, one of them from
another. Agayne, the tymes of the
yeere are not a-lyke through all
the world, but when it is Sommer
to vs, it is winter to some other, and
when it is Spring-tyme with vs, it is
Sommer in another Countrey;
and when it is Haruest with vs, o-
ther people haue Sommer: so whē
it is winter with vs, some Nations
haue Sommer, yea, when Spring
time beginneth with vs, it is Haruest

ueſt in ſome Countryes, and in other Countryes it is Midſommer at the ſame time : but when it is Midſommer with vs, it is Harueſt no where in the world, but middle Winter it is then in two dyuers parts of the world.

And thoſe people whoſe *Zenith* is within 23. degrees and a halfe of any of the Poles, haue their ſhaddowes running round about them: and the neerer they dwell vnder the Pole, the longer is their day, and therefore dooth their ſhaddowes run the oftner about them; for where the day is but 24. howers long, there the ſhaddowes runneth but once about : and where it is

halfe

halfe a yeere long, there it runneth about 103. tymes, and in all other meane places accordingly ; so that those people that haue these shaddowes thus running about them, vnder the North Pole. Then they that dwell vnder the South Pole haue no shaddowes at all, for it is continuall darknes with them: and yet doe they not want lyght although they lacke the Sunne, but only halfe a moneth together, when the Moone is in that halfe of the Zodiake which is out of their Horizont. And though the Sunne and Moone be out of theyr sight, ye see with vs, that we haue a light before Sunne rysing, and after the Sunne

set-

setting: so haue they such a light by the beames of the Sunne, 50. dayes continually, after they haue lost the sight of the Sunne, and so haue they like light 50. dayes together before the Sunne dooth rise to them. And when the Sunne is at the highest with vs, it is at the lowest with diuers other Nations, namely, to all them that dwell vnder the Equinoctiall directly, or South from it; And therefore all those Nations haue Mid-winter when we haue Midsommer.

Nowe followeth, how to erect the Sphere.

First, for the vse of the same, you must place and set your Sphere leuill, that it may stand vpright, and by the needle in the compas in the foote thereof, let it be placed due North and South, then shall the Articke Circle stand North and the Antarticke South.

The next Rule, is to find out the eleuation of the Pole of the heaue͂, in that place wher you mind to obserue the Sphere for, & this being knowne, then turne your Meridian

an Circle, and rectifie the Pole of the Sphere, so many degrees aboue his Horizont, as the Pole of the heauen is eleuated, in the place where you will obserue the same. Then marke the degree of any signe that the Sunne is in that day, whose quantitie you desire to know: set that degree iust in the Horizonte towards the East, and marke what degree of the Equinoctiall is in the Horizonte at the same time: then turne the Sphere West-ward, till the degree of the Sunne bee iust in the Horizonte againe in the West part, and marke then what degree of the Equinoctiall dooth lyght on the Horizont in the East part, accompting truely howe many degrees bee betwixt those two degrees, which you haue marked, and that Arke of the Equinoctiall, is called the Arke of that day: which you may easily turne into howers, accompting 15. degrees to an hower, and for euery degree lesse than 15. accompting 4. minutes of an hower.

Example.

I set the Globe to the eleuation of 52. degrees, and consider the place of the Sunne, the 14. day of August, and find it to be by the Ephemerides, in the first beginning of of *Virgo*, therefore doe I set the beginning of *Virgo* in the very Horizonte, and then doe I see with it, the 137. degree of Equinoctiall in the same Horizonte, which I doe mark; afterward I turne the Sphere til the place of the Sunne, be in the Horizont on the West part, & thē in the East part I mark the place of the Equinoctiall, which is 347. degrees, now abating 137. out of 347. there resteth the whole day Arke, which is 210. degrees, which maketh 14. howers: wherefore I conclude, that the night is but 10 howers, and both those times maketh iust 24. howers.

A 4   *An*

An other way to find the same, more easier.

Example.

For London, the Pole of heauen being rayfed there, 51. degrees and 34. minutes.

Turne your Meridian Circle 51. degrees & 34. minutes, thē the Pole of your Sphere is eleuated to the Latitude of London, thus being finished.

The next Rule, to knowe the day of the month you will practife on.

That is to be found out in the Horizonte

Horizonte Circle of the Sphere, where you shall find also, what degree the Sunne occupieth in the signe that day. Then turne the Circle of the Sunne, that the middle body of the Sunne be brought right against the said degree, in the Zodiake : then turne the whole Globe about West-ward, till the body of the Sunne bee right vnder the Meridian Circle, and there let him stay, till you haue remoued the index of the howerly Circle or Diall, precisely on 12. of the clocke at noone : then turne & bring the Globe backe agayne, to the East part of the Horizonte, where you first found out what degree of the
signe

figne the Sunne was in that day;
Then looke vpon the Dyall, on the
North parte of the Sphere or
Globe, & ye shal find what hower
the Sunne ryseth: thys being done,
bring him back again West-ward,
toward the Meridian Circle, & it
noteth the place of the eleuation
of the Sunne euery hower, till hee
be at his full height vnder the Meridian Circle, then it is sayd to bee
in the very noone steede, for that
place where you vse the Sphere
for, then turn the Globe or Sphere
frō the Meridian Circle, westward,
and it sheweth the nūber of howers
which he falleth from the Meridian height, till the tyme that he setteth

teth vnder the Horizonte, and the index in the Dyall, will tell you the hower that hee setteth vnder the Horizonte that day.

So thus hauing regard to the former instructions, will tell you the tyme of the length of the dayes & nights, in all places of the world, throughout the whole yeere; Prouided alwayes, that ye erect and set the degrees of the Meridian Circle, to the Latitude of the sayde place, where you meane to make your obseruation.

Yet by the way, I will giue you a Rule touching the Sunnes motion, in his Excentrick Circle.

The

The Excentrick Circle in the Sphere or Globe, beeing narrower on the one side than on the other, and hath his Center distant or deuided frō the Center of the world, and is described in the heauen of the Sunne, imagining a lyne from the Center of the Excentricke to the Center of the Sunne, making a complet reuolution at the proper motion of the Sunne.

In the other heauens, imagining a lyne from the Center of hys Excentrick, to the Center of the *Epicicle*.

The *Epicicle*, is a Circle or little Roundle,

Roundle, fixt in the depth of the Excentrick: in which, the Plannet is fixed, and neere to hys Center is moued Circulerlie.

The *Auge*, is a poynt in the circumference of the Excentrick, neerest vnto the Firmament: or it may bee sayde, that the *Auge* is a poynt farthest distant from the Earth.

Aux, in the Greeke tongue is as much to say, as the greatest Longitude or greatest eleuation from the Earth.

The opposite of the *Auge*, is an other poynt in the circumference
of

of the Excentrick, neerest vnto the Earth, and fartheft diftant from the Firmament.

And you muft heere note and vnderftand, the Sunne is not mooued Regularly in the Zodiake, making fo much by his proper motion in one day, as in the other, becaufe his Reguler motion is in refpect of the Center of his own proper Sphere, or orbe wherein hee is moued, whofe Center is diftant without the Center of the world, towards the parts of *Cancer*: fo that the greater part of his orbe Excentrick, is toward the feptentrionall part, where the Sunne paffing by the feptentrionall fignes, is more

diftant

distant from the Earth, and hath more to goe of hys orbe Excentricke, than beeing in the South signes: for, passing by the North signes, he tarryeth 9. dayes more, to describe the halfe of the Zodiake, than the other halfe toward the South part.

And for this cause, the Sunne is more swifter in his motion (in the Zodiake) one tyme, than another: for his motion in one day in the South signes, shall bee greater than it is in one day in the North signes.

And further it followeth, that the sayd vnequall moouing of the Sun obliquite of the Zodiake, certayne
dayes

dayes of winter, with their nights, are longer, than certayne other of Sommer, with their nights: that is to say, that the day naturall in the winter, dooth surmount that in the Sommer, because the right ascention which aunswereth to one dayes motion of the Sunne, beeing in the South, is greater than the ascention for one dayes mouing, being in the North signes.

Next followeth the placing of the other sixe Plannets, in their true order in the Sphere.

These six Plannets, hauing each of them seuerall Spheres, and theyr moti-

motions also seuerall, and vnlyke in tyme to any other: and therefore they are called vvandering starres. These are carryed round about the world, by the vyolence of the first mouer, in 24. howers, that is, euery day once; yet they keepe their places in their Sphere, and haue their proper motions from West towards East.

The *Moone*, with her heauen or Sphere, by her proper motion, giueth her turne from the West to the East, in 27. dayes and 7. howers, with 45. minutes.

Venus, *Mercury,* and the *Sunne,*

in

in a yeere, which is the space of
365. dayes, with 5. howers and 49.
minutes.

Mars, in two yeeres. *Jupiter*, in
12. yeeres. *Saturne*, in 30. yeeres.

The eyght heauen, which is the
Firmament or starry heauen, by his
own proper motion is moued by
the ninth heauen, vpon the beginning of *Aries* and *Libra*, and vpon
these two poynts, accomplisheth
hys Reuolution in seauen thousand yeeres.

This motion is called *Motus trepedationis*, (that is to say) the trembling motion, of *Acceſſ* or *receſſ*.

To

*To rectifie the 6. Plannets, to goe in
theyr due courses about
the Sphere.*

Example.

Saturne, who is the flowest in motion of all the 7. Plannets, the 30. day of Iune, 1592. he is found by the Ephemerides at noone, in his middle motion, to occupie the 16.degree, & 36.minutes of *Cancer*: then I turne the Sphere of *Saturne*, that the body or middle part of his starre or Caracter, be iust vnder the same degree in the signe noted in the Zodiake, then mouing or turning

ning the whole Globe about, from East to the West, sheweth not onely the howers of his rysing and setting, with his Longitudes & Latitudes, but also, what part of the heauens he occupyeth, euery hower of the day & night, as by the Diall it appeareth in the Globe, if it bee set according to the true place of the Sunne.

Lykewise, *Iupiter* is to be sought out, what degree of the signe hee occupyeth the same day at noone, & obserue his Sphere in the same order.

Mars, the lyke; The *Sunne*, I haue shewed you the order before. Then *Venus* followeth, and *Mercury*,

cury, to be ordered and set each of them in theyr seuerall Spheres, then shall appeare euery one of them in theyr seuerall courses, euerie moment of the day: as by the example of *Saturne* before is shewed.

Nowe to the Sphere of the *Moone*, whose motion in her Sphere is neerest to the Earth; the obseruation of her, followeth.

The *Moone*, swifter in course than any of the other Plannets, maketh her reuolution through the twelue signes, twelue times in a yeere oftner than the Sunne. And she is to be placed in lyke wise, according

cording to the former Rules, in the figne fhee is in at noone, that day and hower which you will obferue the Globe or Sphere for, and for euery hower after, adde to her 30. minutes: and (without any great errour) fhee noteth vnto you, her ryfing and fetting, the hower and place of the heauen, euery day and hower where fhe is : the chaunge, quarters, and full Moone, the ebbes and floods, euery hower throughout all the whole day and moneth: and fo confequently throughout the whole yeere, without any great error; Alfo, the depriuing of her lyght by the Earth, in time of her Eclipfes.

T.

To find out by the Instrument in the Dyall, the age of the Moone, with her chaunge, quarters, and full: her aspects with the Sunne: the ebbes & floods, and other necessary Rules, appertayning to the Arte of Nauigation. &c.

Marke at the Coniunction of the Sunne and Moone, (it is sayde) the chaunge of the Moone, is whē the Sunne and shee meeteth together, and then the Moone taketh her lyght of the Sunne: and when she is runne in her course, 24. howers after the change, it is sayde that she is a day olde; then turne the index

dex of the Moone to the figure of
1. And when she is two dayes old,
turne the index of the Moone to
the figure of 2. and so proceede, till
she come to the figure of 7. and
then it is sayd to be in the first
quarter of the Moone: then at the
figure of 15. shee is in opposition
with the Sunne, then it is said to be
a full Moone: then shee gathereth
euery day in her decrease, towards
the Sunne. And when she is 7.
dayes past the full, then she hath
lost half her roundnes of her light,
and is sayd to be last quarter; and
so gathereth euery day, neerer and
neerer the Sunne, till she be depry-
ued quite from her light, and then

it

it is sayd to be at chaunge agayne, and a newe Moone: and after her chaunge, then her lyght begins to increase agayne, euery 24. howers, 48. minutes; which yeeldeth in 15. dayes, 12 howers. And so much she is iust of the Sunne, at the time of the full Moone. And if you will marke the distances, betwixt the index of the Sunne, which poynts the howers in the Dyall, and in the index of the Moone: you shal find alwayes what distance the Sunne and Moone are a sunder; thys is called amongst the Marrines, the shifting of the Sunne and the Moone, & hereby they shall know theyr ebbes and floods, as appeareth

reth in the vttermost part of the Dyall : and also, the 32. poynts of the Compas sheweth the same.

Another Rule for the hower of the two starres afore-sayde, called the Guardes, and of some, called Charles Wayne, or Charles Carte : likning fower starres to fower wheeles, and the other three starres to three Oxen.

And the first starres I take for my purpose, and declare at euerie monethes end, at what hower they are full West, and the howers that they are West and by North, and North-west, and North-west and by North, and full North; and so
round

round about 24. howers.

Example.

JANUARY.

From the 2. day of *January*, to the 17. they are North-east, at 5. at night: so you must turne the former Guarde, that the thrid going from the same to the North starre, may fall iust vppon the North North-east poynt of the Compas, at 5. of the clocke at night: and then turn the Globe round about, and it noteth euery hower, and euery poynt in the Compas, that coast where they are situated, and
al-

alſo, what hower they riſe and ſet vnder the Pole. And by this example put, you may proceede throughout euery moneth in the yeere, according to the tymes of theyr beeing, and euery hower in the ſame. As at 6. of the clocke at night, they are North-eaſt and by North.

at 7. North-eaſt.
at 8. Eaſt and by North.
at 9. Full Eaſt.
at 10. Eaſt and by South.
at 11. South-eaſt.
at midnight, South-eaſt and by South.
at 1. South-eaſt.
at 2. South-eaſt and by South.

at 3.

at 3. Full South.
at 4. South-west and by South.
at 5. South-west.
at 6. South-west and by South.
at 7. South-west.

From the 17. to the last, they are North-east and by North at 5. at night.
at 6. North-east.
at 7. East and by North,
at 8. Full East.
at 9. East and by South.
at 10. South-east
at 11. South-east and by South
at midnight, South South-east
at 1. South-east and by South
at 2. Full South
at 3. South-west and by South

at 4.

at 4. South-west,
at 5. South-west and by South
at 6. South-west
at 7. West and by South.

From the last to the 15. of *February*, they are North-east at 5. at after noone.

at 6. East and by North
at 7. Full East.
at 8. East and by South
at 9. South-east
at 10. South-east and by South
at 11. South-east
at midnight, South South-east and by South.
at 1. Full South
at 2. South South-west and by South.

at 3.

at 3. South-west
at 4. South-west and by South
at 5. South-west
at 6. West and by South
at 7. Full West

FEBRUARY.

*The howers of the two starres of
Charles Wayne.*

From the 15. to the 1. of *March,* they are full East, at 6. at after noone.
at 7. East and by South.
at 8. South-east
at 9. South-east and by south
at 10. South-east

at 11

at 11. South-east and by south
at midnight, full South
at 1. South-west and by south
at 2. South-west
at 3. South-west and by south
at 4. South-west
at 5. West and by South
at 6 Full West.

From the 1. of *March* to the 16. of *March*, they are East and by South, at 6. at after noone.
at 7. South-east
at 8. South-east and by south
at 9. South-east
at 10. South-east and by south
at 11. Full South
at midnight, South south-west and by south.

at 1.

at 1. South south-west
at 2. South-west and by south
at 3. South-west
at 4. West and by South
at 5. Full West
at 6. West and by North

MARCH.

The howers of the two starres of Charles Wayne.

From the 16. to the 1. of *Aprill*, they are South-east and by south, at 7. at after noone.
at 8. South South-east
at 9. South south-east & by south
at 10. Full South
at 11 South southwest & by south

at mid-

at midnight, Southwest
at 1. South-west and by south
at 2. South-west
at 3. West and by South
at 4. Full West
at 5. West and by North

 From the 1. of *Aprill*, to the 16. they are South South-east, at 7. at after noone.
at 8. South south-east & by south
at 9. Full South.
at 10. South southwest & by south
at 11. South south-west
at 12. South-west and by south
at 1. South-west
at 2. West and by south
at 3. Full West
at 4. West and by North

at 5.

at 5. North-west

APRILL

*The howers of the two starres of
Charles Wayne.*

From the 16. of *Aprill* to the 2. of
May, they are full South, at 8. at after noone.
at 9. South southwest & by south
at 10. South south-west
at 11. South-west and by south
at midnight, South-west.
at 1. West and by South
at 2. Full West.
at 3. West and by North
at 4. North-west

From the 2. of *May* to the 18.

they

they are South South-west and by South, at 8. at after noone.
at 9. South South-west
at 10. South-west and by South
at 11. South-west
at midnight, West and by South
at 1. Full West.
at 2. West and by North
at 3. North-west
at 4. North-west and by North.

MAY.

The howers of the two starres of Charles Wayne.

From the 18 of *May* to the 2. of *June*, they are South-west and by South, at 9. at after noone.

at 10.

at 10. South-west
at 11. West and by South
at midnight, Full West
at 1. West and by North
at 2. North-west
at 3. North-west and by North
From the 2. day of *June*, to the 18. they are South-west, at nine at after noone.
at 10. West and by south
at 11. Full West
at 12. West and by north
at 1. North-west
at 2. North-west and by north
at 3. North-west

E 5   *JUNE*

IVNE.

The howers of the two starres of Charles Wayne.

From the 18. of *Iune*, to the 4. of *Iuly*, they are West and by South, at
at 9. at after noone.
at 10. Full West
at 11. West and by north
at midnight, North-west
at 1. North-west and by north
at 2. North north-west
at 3. North northwest & by north

From the 4. of *Iuly* to the 20. they are full West at 9. at after noone.
at 10. West and by north
at 11. North-west

at mid-

at midnight, North-west and by north.
at 1. North north-west
at 2. North north-west and by north.
at 3. Full North

JULY.

The howers of the two starres of Charles Wayne.

From the 20. of *July*, to the 4. of *August*, they are full West, at 8. at night.
at 9. West and by north
at 10. North-west
at 11. North-west and by north
at midnight, North-west

at 1. North north-west and by north.

at 2. Full North

at 3. North northeast & by north
From the 4. of *August* to the 20. they are West and by north, at 8. at after noone.

at 9. North-west

at 10. North-west and by north

at 11. North-west

at midnight, North-west and by north.

at 1. Full North

at 2. North north-east and by north.

at 3. North north-east

at 4. North-east and by north

AUGUST.

The howers of the two starres of Charles Wayne.

From the 20. of *August*, to the 4. of *September*, they are North-west, at 8. at after noone.

at 9. North-west and by north
at 10. North north-west
at 11. North north-west and by north.
at midnight, full north
at 1. North northeast & by north
at 2. North north-east
at 3. North-east and by north
at 4. North-east

From the 4. of *September* to the 19. they

19 they are North-weſt at 8. at after noone.
at 9. North north-weſt
at 10. North north-vveſt and by north.
at 11. Full north
at midnight, North north-eaſt and by north
at 1. North north-eaſt
at 2. North-eaſt and by north
at 3. North-eaſt
at 4. Eaſt and by north

SEPTEMBER.

The howers of the two ſtarres of Charles Wayne.
From the 19. of *September* to the 5. of

5. of *October*, they are North-west and by north, at 7. at after noone.
at 8. North north-west
at 9. North northwest & by north
at 10. Full north
at 11. North northeast & by north
at midnight, North-east
at 1. North-east and by north
at 2. North-east
at 3. East and by north
at 4. Full East
at 5. East and by South

From the 5 of *October* to the 20. they are North north-west, at 7. at after noone.
at 8. North north-west and by north.
at 9. Full north

at 10.

at 10. North northeast & by north
at 11. North north-east
at midnight, north-east & by north
at 1. North-east
at 2. East and by north
at 3. Full East
at 4. East and by South
at 5. South-east

OCTOBER.

The howers of the two starres of
Charles Wayne.

From the 20 of *October*, to the 3. of *Nouember*, they are North north west at 6 at after noone.

at 7. North north-west and by north

at 8.

at 8. Full north
at 9, North northeast & by north
at 10, North north-east
at 11, Northeast and by north
at midnight, north-east
at 1, East and by north
at 2, Full East
at 3, East and by South
at 4, South-east
at 5, South-east and by south
at 6, South south-east

From the 3, of *Nouember*, to the 18, they are North northwest & by north, at 6, at after noone.
at 7, Full North
at 8, North northeast & by north
at 9, North north east
at 10, North east and by north

at 11,

at 11, North-east
at midnight, East and by north
at 1, Full East
at 2, East and by South
at 3, South-east
at 4, South-east and by south
at 5, South-east
at 6, South south-east & by south

NOVEMBER.

The howers of the two starres of
Charles Wayne.

From the 18. to the 3, of *December*, they are full North, at 6, at afternoone.
at 7, North northeast & by north
at 8, North north-east

at 9,

at 9, North-east and by north
at 10. North-east
at 11. East and by North,
at midnight, Full East.
at 1. East and by South.
at 2. South-east
at 3. South-east and by South
at 4. South-east
at 5. South South-east and by South.
at 6. Full South
at 7. South South-west and by South

From the 3. of *December*, to the 17, they are full North, at 5, at after noone.
at 6. North north-east & by north
at 7. North north-east.

at 8.

at 8. North-east and by north
at 9. North-east
at 10. East and by North.
at 11. Full East.
at midnight, East and by South.
at 1. South-east.
at 2. South-east and by South.
at 3. South South-east.
at 4. South Southeast & by south
at 5. Full South
at 6. South southwest & by south
at 7. South South-west

DECEMBER.

*The howers of the two starres of
Charles Wayne.*
From the 17. of *Decmber*, to the
1. of

1. of *January*, they are full North,
at 4. at night.
at 5. North northeast & by north
at 6. North north-east
at 7. North-east and by north
at 8. North-east
at 9. East and by North
at 10. Full East.
at 11. East and by South
at midnight, South-east
at 1. South-east and by south
at 2. South South-east
at 3. South south-east & by south
at 4. Full South.
at 5. South Southwest & by south
at 6. South south-west
at 7. South-west and by South
From the 1. of *January* to the 16.
F 2   they

they are North North-east at 5. at
after noone.
at 6. North-east and by north
at 7. North-east
at 8. East and by north
at 9. Full East
at 10. East and by South.
at 11. South-east
at midnight, Southeast & by south
at 1. South South-east
at 2. South south-east & by south
at 3. Full South
at 4. South South-west and by South.
at 5. South South-west
at 6. South-west and by south
at 7. South-west.

A ready note in fewe words, for the difference of howers, according to the distance myles, from East to West vnder the Equinoctiall.

FIRST, you shall vnderstand, that 15. myles difference from East toward West, doth make the Sunne rysing, the noone steed, and Sunne setting, to be later by one minute of an hower: & so 30 miles, 2.minutes: 120. myles, 8. minutes: 225. myles, 15. minutes: which is a quarter of an hower. And he that is ready in accompt of Arithmatique, may find it out by the Rule of proportion. As for Example.

London hath Latitude 51. degree

F 2 and

and 30. minutes, or there abouts, I trauayle East-wards from London 2000. myles; My desire is, to know the difference of theyr Longitudes, and the time of theyr noone steeds, for when it is 12. of the clock with vs at London, 2 0 0 0. myles East-ward from London, is then but 2. of the clock and 13. minutes at after noone. And 2000. myles West from London, it was then with them, but 10. of the clock and 13. minutes in the fore noone: the difference of these 3. places, one from the other, is to bee founde by the Rules in Arithmatique, as followeth.

If 15. myles in Longitude, East-ward

ward from London giue one minute of time, what gyueth 2000. myles.

myles, 15. ⟍ 1. minute of time.
myles, 2000 ⟋ 133. (5.

I worke it in this manner, deuiding 2000 miles by 15 myles, and it yeeldeth 133. times 15. myles, and 1. third part of 15. miles, to be deuided into 15 parts. Now, alowe to euery 15. myles, one minute of tyme, (as you haue heard me say before) and reduce them into howers, in thys manner as followeth.

60. minutes maketh a degree of the

F 3 Equi-

Equinoctial; so then deuide 13⁚.by 60. minutes of time, & the Cotient will be 2. and 13.will remayne: that is, 2. howers and 13. minutes difference, betweene that place & London. And in this wise may you worke by the Rules in Arithmatique, to find the West Longitude from London.

There bee some persons that make a great obscuritie, in finding out the Longitudes in sayling East and West; a thing once knowne, & of no great importance, as ready to bee found out as the Latitudes. A little Briefe (therfore) I will giue you, to vnderstand the same skill.

Let the Marriner, Sayler, or other persons,

persons, prouide him a perfect Watch, or Clock, arteficially made by a Clock-maker. Let him set the same by the hower of the day in that place you are in, and to come by the true place of the Sunne, your Astrolob quadrant crosse staffe, or other Instrument, will serue you to take the heigth of the Sunne, & to find out the true hower in euery seuerall Latitude, with the helpe of the Rules before. And the true hower beeing found of the day in this manner, sette your Clocke or Watch. Then trauell either by Sea or Land, and when you are 40. myles, or 60. more or lesse, distant of the place you went from, then

F 4   looke

looke to your Clocke or Watch, howe many howers haue paſſed ſince you ſet on your iourney: then take your Quadrant or orther Inſtrument, & take the heigth of the Sunne in that place you are in; and if the time of the day taken with your Inſtrument, doe agree with your clock, be you ſure your place is North or South, from the place you came from, and therefore haue the ſame Longitude and Meridian lyne; But if the tyme differ, ſubſtract the one out of the other, and the difference turne in degrees and minutes of the Equinoctiall: and 15. minutes of the Equinoctiall, maketh one minute of time, according vnto

vnto the Rules going before, you may thus knowe the Longitudes, difference of howers, and times, betwixt any two places East or West ward. The Latitudes are easily found out by the Altitudes of the North or South Poles, and also, by the Meridian heigth of the Sunne at noone.

Example.

I find the Sunne to possesse the 1. minute of *Geminie:* his heigth in the Meridian lyne at noone with vs heere at London, is 54, degrees and 10, minutes, and his declination is 15, degrees and 40, minutes:

now

nowe I substract the declination out of the Sunnes heigth, & there remayneth 39, degrees and 30, minutes, this I take from 90, degrees, and I find the place in height 51, degrees and 30, minutes; And this you must beare in memory, that if the Sunne haue South declinatiō, you must adde it to the sayde Altitude : then adding or substracting that number from 90, degrees, ther shall remaine the true eleuation of the Pole.

Alwayes beare this in memorie, if the Sunne be on the South side of the Equinoctiall lyne, it is called South declination; if on the North side, North delination; And thys
hath

hath a most singuler vse in the Arte of Nauigation, and by it you may finde out the heigth of the Pole in all places wher you trauell.

¶ *Heere followeth the* Degrees, called *Putei*, *Fortune*, *Lucidi*, *Tenebrosi*, *Vacui*, *Masculini*, and *Feminei*, *Fumosi* and *Azamene*, *in all the* 12. *Signes of the Zodiake. By this letter* p. *is ment Putei: for. signifieth Fortune: and a. Azamene: this letter* l. *betokeneth Lucidi:* t. *Tenebrosi: this letter* v. *noteth Vacui: and* m. *Masculini: this letter* f. *signifieth Feminei: and this sillable* fu. *noteth Fumosi.*

Aries

Aries. *Aries.*
Degrees,   Degrees,
1, t. m.   16, t. f. p.
2, t. m.   17, l. f.
3, t. m.   18, l. f.
4, l. m.   19, l. f. for.
5, l. m. p. 20, l. f.
6, l. m.   21, v. f.
7, l. m.   22, v. f.
8, l. m.   23, v. m. p.
9, t. f.   24, v. m.
10, t. f.   25, l. m.
11, t. m. p. 26, l. m.
12, t. m.   27, l. m.
13, t. m.   28, l. m.
14, t. m.   29, l. m. p.
15, t. m.   30, v. m.

Taurus,

Taurus   *Taurus*
Degrees,   Degrees,
1, f. t.   16, v. f.
2, f. t.   17, v. f.
3, f. t. for.   18, v. m
4, f. l.   19, v. m
5, f. l. p.   20, v. m
6, m. l. a.   21, l. m.
7, m. l. a.   22, l. f
8, m. v. a.   23, l. f
9, m. v. a.   24, l. f. p
10, m. v. a.   25, l. m. p
11, m. v.   26, l. m
12, f. v. p.   27, l. m. for
13, f. l.   28, l. m
14, f. l   29, t. m
15, f. l. for   30, t. m

Geomantie

Gemini, *Gemini,*
Degrees, Degrees,
1, l. f. 16, v. m.
2, l. f. p 17, l. f. p
3, l. f 18, l. f.
4, l. f 19, l. f
5, t. f 20, l. f.
6, t. m 21, l. f.
7, t. m 22, l. f.
8, l. m 23, t. m
9, l. m 24, t. m.
10, l. m 25, t. m.
11, l. m. for 26, t. m. p
12, l. m. p 27, t. m
13, v. m 28, v. f
14, v. m 29, v. f
15, v. m 30, v. f. p

Cancer

Cancer.	*Cancer.*
Degrees,	Degrees,
1, l. m. for.	16, v. m
2, l. m. for	17, v. m. p
3, l. f. for	18, v. m
4, l. f. for	19, fu. m
5, l. f	20, fu. m
6, l. f	21, l. m.
7, l. f	22, l. m
8, l. f	23, l. m. p
9, l. m. a	24, l. f.
10, l. m. a	25, l. f
11, l. f. a.	26, l. f. p
12, l. f. a. p	27, l. f.
13, t. m. a	28, l. f
14, t. m. a	29, t. v
15, v. m. a. for	30, t. v. p

G   *Leo.*

Leo.　　　　　*Leo.*
Degrees,　　　Degrees,
1, t. m.　　　16, fu. f
2, t. m. for　 17, fu. f
3, t. m　　　 18, fu. f. a
4, t. m　　　 19, fu. f. for
5, t. m. for　 20, v. f. fu.
6, t. m. p　　21, v. f.
7, t. f. for　 22, v. f. p
8, t. f　　　　23, v. f. p
9, t. m　　　 24, v. m
10, t. m　　　25, v. m. a
11, fu. m　　 26, v. m. a
12, fu. m.　　27, l. m.
13, fu. m. p　28, l. m. p
14, fu. m　　 29, l. m
15, fu. m. p　30, l. m

Virgo,

Virgo,	*Virgo,*
Degrees	Degrees,
1, f. t.	16, f. l. p
2, f. t.	17, f. fu
3, f. t. for	18, f. fu.
4. f. t	19, f. fu.
5, f. t.	20, fu. for
6, f. t	21, m. fu. p
7, f. l	22, m. fu
8, f. l. p	23, m. v
9, m. v	24, m. v
10, m. v	25, m. v. p
11, m. l	26, m. v
12, m. l	27, m. v
13, f. l. p	28, m. t.
14, f. l. for	29, m. t
15, f. l	30, m. t

Libra

Libra,
Degrees
1, l. m. p.
2, l. m. p
3, l. m. for
4, l. m
5, l. m. for
6, t. f
7, t. f. p
8, t. f
9, t. f
10, t. f
11, l. f
12, l. f
13, l. f
14, l. f
15, l. f

Libra
Degrees,
16, l. m.
17, l. m
18, l. m
19, t. m
20, t. m. p
21, t. f. for
22, l. f.
23, l. f
24, l. f
25, l. f
26, l. f
27, l. f
28, v. m
29, v. m
30, v. m. p

Scorpio,

Scorpio,	Scorpio,
Degrees,	Degrees,
1, t. m.	16, l. m
2, t. m	17, l. m
3, t. m	18, l. f. for
4, l. m	19, l. a f
5, l. f. for	20, l. f. for
6, l. a. f	21, fu. f
7, l. f. for	22, fn. f. p
8, l. f.	23, v. f. p
9, v. f. p	24, v. f
10, v, f. p	25, v. m
11, v. f	26, v. m
12, v. f	27, v. m. p
13, v. f	28, t. m
14, v. f	29, t. a. m
15, l. m	30, t. m

Sagitarius

Sagitarius
Degrees,
1, l. m. a
2, l. m
3, l. f
4, l. f
5, l. f
6, l. m
7, l. m. a p
8, l. m. a
9, l. m
10, t. m
11, t. m
12, t. m. p
13, l. f. for
14, l. f
15, l. f. p

Sagitarius
Degrees,
16, l. f
17, l. f
18, l. f. a
19, l. f. a
20, fu. f. for
21, fu. f.
22, fu. f
23, fu. f
24, l. f. p
25, l. m
26, l. m
27, l. m. p
28, l. m
29, l. m
30, l. m. p
Capricor.

Capricor.
Degrees,
1, t. m
2, t. m. p
3, t. m
4, t. m
5, t. m
6, t. m
7, t. m
8, l. m
9, l. m
10, l m
11, fu. m.
12, fu. f. for
13, fu. f. for
14, fu. f. for
15, fu. f

Capricor.
Degrees
16, l. t
17, l. f. p
18, l. f
19, l. f
20, t. m. for
21, t. m
22, t. m. p
23, v. m
24, v. m. p
25, v. m
26, t. m. a
27, t. m. a
28, t. m. a p
29, t. m. a
30, t. m

Aquary

Aquary.
Degrees
1, fu. m. p
2, fu. m
3, fu. m
4, fu. m
5, l. m
6, l. f
7, l. f. for
8, l. f
9, l. f
10, t. f
11, t. f
12, t. f. p
13, t. f
14, l. f
15, l. f

Aquary.
Degrees
16, l. m. for
17, l. m. for
18, l. m. a
19, l. m. a
20, l. m. for
21, l. m
22, l. f. p. v
23, v. f
24, v. f. p
25, v. f
26, l. m
27, l. m
28, l. f
29, l. f. p
30, l. f

Pisces.

Pisces.
Degrees
1, t. m
2, t. m
3, t. m
4, t. m. p
5, t. m
6, t. m
7, l. m
8, l. m
9, l. m. p
10, l. m
11, l. f
12, l. f
13, t. f. for
14, t. f
15, t. f

Pisces.
Degrees
16, t. f.
17, t. f
18, t. f
19, l. f
20, l. f. for
21, l. m
22, l. m
23, v. m
24, v. f. p.
25, v. f
26, l. f
27, l. f. for. p
28, l. f. p
29. t. m.
30. t. m.

To

To know the place of the Sunne, by the Rule of memorie: And to knowe in what degree the Sunne is, without respect of minutes.

Beare in memorie these numbers that heere-after followeth. 11. 10. 11. 10. 11. 12. 13. 14. 13. 14. 12. 12.

The first 10. standeth for *Ianuary*, the second for *February*, with their signes, & so the rest; And to know in what degree the Sun is, you shal take away the dayes that are applied to euery moneth, according to the said numbers of the dayes, for the which you desire to know the place of the Sunne, and in them that remayn, in so many degrees is the Sunne, of the signe into which

it

it entereth that moneth. And if the
dayes paſt of the moneth, ſhall bee
leſſe then the dayes applyed to the
ſame moneth, you ſhall ioyne 30.
with thoſe dayes paſt of the moneth, & of the Sunne that amoun-
teth, you ſhall take away the daies
applyed to the ſayde moneth, and
the reſt ſhall be the degrees, in the
which the Sunne ſhall be, of the
ſigne of the moneth paſt. As for
example.

Moneth.

Moneth,	Degrees,	Signes.
Ia.	11.	Aquarius
Fe.	10.	Pisces.
Ma.	11.	Aries
Ap.	10.	Taurus
Mai.	11.	Gemini
Iune.	12.	Cancer
Iuly.	13.	Leo
Au.	14.	Virgo
Sep.	13.	Libra
Octob.	14.	Scorpio
No.	12.	Sagitarius
De.	12.	Capricornus

The 12. day of *October*, taking away 14. that were applyed, remaineth 28. degrees of *Scorpio*, where the Sunne is.

An

Another Example.

The sixt of *December*, which are lesser then 12. which is applied vnto it, if we ioyn 6. to 30. which are the daies of the moneth next afore, they make 36. and from them wee take away the 12. & there rest 24. degrees is the Sunne of the month before, which is *Sagitarius*.

A Rule to know when the Sunne entreth into euery of the 12. Signes.

And that wee may in the yeeres to come, know the day, hower, and minute, in the which the Sun entreth

treth into euery signe, we will follow this order; vpon the day, howers, & minutes, that the Sunne entreth into euery signe, the yeere 1545. we must add for euery yeere 5. howers and 49. minutes, which with the 365. dayes which euery yeere cōtaineth, shalbe the time in the which the Sun accomplisheth his reuolution. And because that in the yeere of the Bebysextile or Leap-yere, is added to *February*, 1. day more to his 28, which we haue once in 4. yeeres, from 6. to 6. howers; if we shal take from the Computation, that we haue giuen turning one day backward, as shall be in the yeere 1548. and vppon that

remayremaineth shall return in the yeere following, of 1549 to add 5, howers 49. minutes, and as much more euery other yeere following shall be a certaine Rule for euer.

And it is to be noted, that the degrees and minutes which we haue touched before, are properly for the citty of *Cadiz*. And if we desire to apply thē for other ciries or places more Eastward, thē for euery 15 degrees that they are distant from *Cadiz*, in Longitude, we must add one hower, And if for the Cities or places more westward in like manner, for euery 15. degrees, we must take away one hower, by reason of the course of the Sun, by his Rapte
mouing

mouing frō the East to the West. For it is certain, that when with vs in *Cadiz* it is 12. howers of the clock to them that are 15. degrees Eastwarde from vs, it is one of the clocke : and to them that are from vs 15. degrees towards the West, it is 11. of the clocke. And thus may you apply it to euery seueral Longitude East or West, gyuing to euery 15. miles, one minute of tyme according to the other Rules in Arithmatique specified.

FINIS.

AT LONDON
Printed by Iohn Charlwood.